THOSE ARE JUST CARTOONS

EARTH IS NOT A BALL

ARCTURO TATE

Chapter 1
Rationality

You think it’s so ridiculous?
the idea of flat-earth?

Your criticisms are empty
You fear debate
You realize you have no rational argument for believing in the ball
Those are just cartoons

This is not about one philosophy vs. another
“Scientific” authorities are lying to you
The same way they lied to your parents and grandparents
With a globe in every classroom, every movie, every advertisement

You are under a spell
But spells don’t work without the believer
You imprison yourself when you defend these lies

Science isn't ever settled

Refusal to acknowledge arguments that contradict your ideas
is un-scientific verily

We who dare question scientismic dogma
are disregarded and labeled as anti-science

Inversion of facts
and projection of your own transgressions upon your enemy
are trademarks of the most dangerous religions

You are programmed from an early age to believe
earth is a spinning ball floating through outer space

There is no real proof to support this belief
On the contrary ball-earth is easily disproven

You believe those are real photos – the cartoons produced by nasa
You believe in the fantasies of your scientismic priests
about outer space and the origins of the world
Yet they never advance beyond theory

The act of believing in anything is inherently un-scientific

You are trained to mark flat-earthers as stupid
But if you'd open your mind a bit
you'd see that flat-earthers are the true scientists
blowing the whistle on the ball-earth and nasa hoaxes

Flat-earthers use measurement, common sense, logic
Flat-earthers do not Believe

A Comparison between Ball-earthers and Flat-earthers

A flat-earther can show you evidence of flat-earth

Sunset trail
Stand at a body of water with the sun above the horizon
The light reflects off the water in the shape of a straight trail
This is only possible since earth is flat

Over the curve
Objects viewed across water can be seen
farther than the ball-earth horizon would allow (see Chapter 3)

From a mountain the horizon is flat
From an airplane. From a weather balloon. Flat horizons

The horizon rises with the observer
The horizon is at eye-level out the windows of both sides of the airplane
So earth can't be a ball

There's been no change in the stars for millennia
So earth isn't zipping through the galaxy

Earth wind is never 1000 mph
So earth isn't spinning
$Earth\ Circumference = 25{,}000\ miles$
$24\ hours$ in a ball-earth rotation
$25{,}000\ miles \div 24\ hours > 1{,}000\ mph$ at sea level at the equator

On the left hand
a ball-earther has no evidence to show

Ball-earther empirical confirmations and measurements
can be explained just as well or better using the flat-earth model

The only argument you're left with is faith in your priests

Do you really think those cartoons are photos? Cartoons aren't evidence
If nasa and satellites and the ISS were legit
there would be millions of irrefutable evidential photos and videos

Nasa makes many blaring movie mistakes
How can you ignore the air bubbles during space walks?
And the cable harnesses and the hair mousse and green screens?

The purpose of the moon landing hoax
was to convince you the moon is made of rock and dirt

How come there's no blast crater under the lunar lander?
Who shot the video of it taking-off earthbound?
How could it have made such an arduous journey
built of butcher paper and shower curtain rods?
How can moonwalkers phone the president from 250,000 MILES away?

Flat-earth allows for travel amongst the stars and secret space programs
But rocks in the sky and orbiting satellites are rubbish
Modern GPS and cell phones are facilitated with ground-based antennas
Satellites are just cartoons

If saturn is a ball, it doesn't mean earth is a ball
Wandering stars are not in the same category with land and oceans

You pledge allegiance to the heliocentrists
Because why would they be lying
But that's not evidence either

Computer graphic art continues to improve
Soon you might not be able to tell what is a cartoon
But time is not to that point

Nasa aren't innocent creative artists
These are top-level scientific authorities
engaged in criminal mass-brainwashing
government-subsidized/taxpayer-funded

Some of the cartoons are pretty realistic
But don't let them control your mind with it
They refute themselves enough to warrant your suspicion
Be rational

If ball pushers had real proof of the ball
why wouldn't they show you? Why just show you cartoons?
Do they think you're too dumb to understand?

Children tend to trust authority figures
Teachers are supposed to teach truth

If you were a grown up with critical thinking skills
when initially you were told about heliocentrism and spaceballs
then you would know right away it is nonsense

Flat-earthers graduated from the same indoctrination programs as you did
Flat-earthers were once just as convinced of the ball

Flat-earthers refuse to stay asleep
Flat-earthers hold the possibility
that the government is lying on a massive scale

Chapter 2

The Great Deception

Scientism is a major arm in the eternal battle between good and evil

Evil conspiracies are unpopular dinner-table topics
The perceived insanity is a wake-up call
And you keep hitting the snooze

The word evil turns you away
Religious ideas are widely ridiculed and escaped
You brandish science like a crucifix to a vampire
And you are righteous in principle

However most of what you call science
is backed only by theory and authority and lies
No evidence. No reality

It is actually an evil religion called scientism

They deceive you about the most important subjects in life
They poison your heart and mind and body
They feed on your energies from cradle to grave

Scientism has infiltrated every sector of society
Public policy is drawn from it
And its religiosity slithers by unrecognized

Everybody throughout history knew earth is flat
up until the mid-1800's, a period known as the masonic renaissance
when free-masons aggressively accelerated the scientism agenda

Ball-earth is not a symbol of the advancement of knowledge
but instead a symbol of enslavement and broken trust

Scientismists/communists have created an entire fake society
Government is fake. News is fake. Education is fake.

The movies are psychological warfare
Banksters manifest riches out of thin air
Courts protect criminals
Doctors harm patients
Governments tax carbon dioxide

You're a cave prisoner mesmerized by Plato's puppet show
You actually believe ball-earth is a reality
The world is upside-down

You are given the globe before you can walk
Ball-earth in every classroom and TV program
You are saturated with ball-earth, and still as an adult
Even most university professors believe in the ball

Large media networks, including "scientific" journals
are run by people getting paid big money to deceive you
They are the ones being evicted from their thrones by the truth
You can't rely on them to report the current events
much less the corruption of engrained scientism

The term conspiracy theorist was coined by the CIA
in 1964 directly after the Kennedy assassination
to ridicule anybody who asks questions

Citizens are trained to stifle each other
Trained to attack anyone who speaks against the mainline

It is naïve to think conspiracies don't exist
And exposing lies doesn't require theory

The world is run by scientismists/fascists/satanists
Their darkness upon the world protects their operation

Such grand liars rely on you believing their crimes are impossible
The more incredible the crime, the less risk of punishment
and the higher praise they receive from their masters
Scum rises to the top

The superior motive for their crime is the crime itself
Satanists commit crimes because that's when they feel their satan

They seek to destroy anything good
They actively keep you stupid and sick
They conduct inhumane science experiments
They clash civilizations and profit from wars
They intentionally spoil the air and water and farmlands
They openly plot to exterminate the majority of the population
They employ pedophile kidnappers
They worship devils

Murdering humanity throughout history and pre-history
THEY are the ones who proclaim
earth is a spinning ball flying through outer space

Chapter 3

Math Doesn't Lie

Ball-earth Math: How far is the horizon?

Don't fear the math. It won't hurt

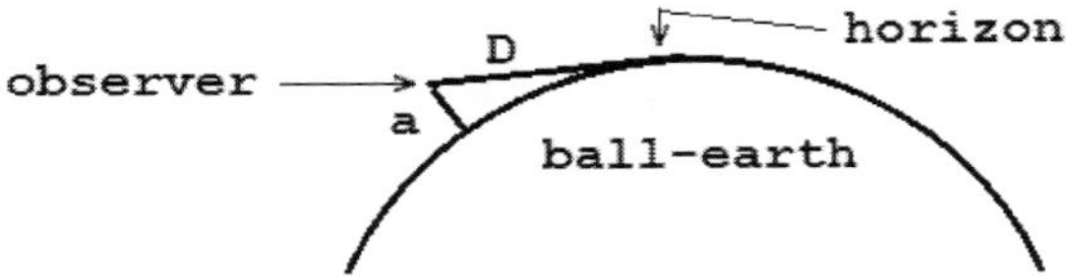

$a = \frac{2}{3}D^2$

$a = \frac{2}{3}D^2$ is a ball-earth equation. Flat-earthers didn't come up with it
It's based on earth ball circumference of 25,000 miles (see Appendix B)
a is altitude {feet} above sea level of the observer
D is the Distance {miles} to the horizon

Example 1
Standing at the edge of the sea
$a = 6\,ft$ eyeheight
$D = \sqrt{\frac{3}{2}(\,6ft\,)} = 3\,miles$ 3 miles to the horizon

Example 2
Standing on the deck of a ship
$a = 20\,ft$ eyeheight
$D = \sqrt{\frac{3}{2}(\,20ft\,)} = 5.5\,miles$ 5.5 miles to the horizon

5.5 miles?
Men at sea should only be able to see 5.5 miles?
Ball-earthers are lying to you
Men at sea know they can see farther than 5.5 miles
Even without a telescope

With a telescope they can see as clear as can be
Objects go into the horizon not over
Earth is flat

The Chicago News

…An interesting phenomena on this clear night
Chicago is visible from across Lake Michigan
60 miles away. That's impossible some say

Scientists are calling it a superior mirage
Light reflects off the water vapor
and we can see what's on the other side of the horizon…

Road apples
They're totally making this up about the superior mirage
The buildings are seen right-side up
That's not how mirages work

How high would the buildings have to be
to be visible from 60 miles?

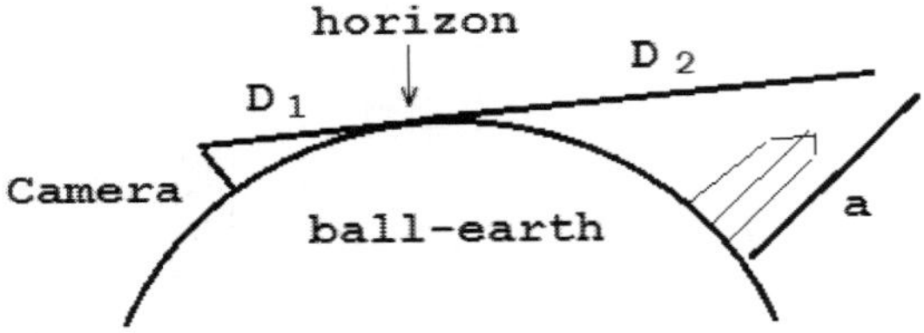

Camera is at the edge of the lake and sears tower across the lake
$D_1 + D_2 = 60\ miles$

$camera\ height = 20\ ft$

$D_1 = \sqrt{\frac{3}{2}(\ 20ft\ )} = 5.5\ miles$ from camera to the horizon

$D_2 = \mathbf{60\ miles} - D_1 = 54.5\ miles$ from horizon to Chicago

$a = \frac{2}{3}(\ 54.5^2\ ) = 1{,}980\ ft$ height required to view from 60 miles
And it wouldn't be a good view
The point you'd be able to see is just the tip of the building
And it would be tilted away from you

In reality how high is the sears tower? $1{,}450\ ft$
But $1{,}450\ ft < 1{,}980\ ft$
So if earth is a ball the buildings should be hidden
by a few hundred feet of marine horizon

Not only that
The camera catches the buildings all the way to the ground
And they're not tilted away

It looks like somebody's ball-earth theory
is full of 2000 ft gaping holes
Earth is flat

Galileo (free-mason) uses the telescope
to cement heliocentricism into society's mind
Ironically the best proof that earth isn't a ball
is viewing objects across the sea with a telescopic camera

All around the world
islands are viewable from 100+ miles off neighboring coasts
This is impossible if earth is a ball
This is proof that earth is flat

The island of Elba can be viewed from Genoa, Italy
The real, measured distance from Genoa to Elba is 125 miles
The peak altitude of the island of Elba is 3,340 ft

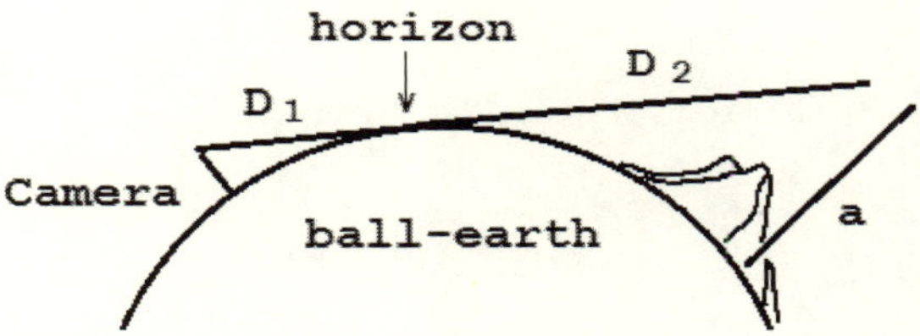

$camera\ height = 70\ ft$ from a hilltop in Genoa

$D_1 = \sqrt{\frac{3}{2}(\ 70ft\ )} = 10.2\ miles$ from camera to the horizon

$D_2 = \mathbf{125\ miles} - D_1 = 114.8\ miles$ from horizon to Elba

$a = \frac{2}{3}(\ 114.8^2\ ) = 8{,}786\ ft$

If earth is a ball
and you're trying to view an object 125 miles away
the horizon completely blocks anything shorter than 8,786 ft

But Elba can be seen from Genoa
And the altitude of Elba is only 3,340 ft
So earth is flat

Statue of Liberty in NYC
$a = 326\, ft$ altitude above sea level

If earth is a ball
what is the farthest distance away you could be
while still being able to see the statue?

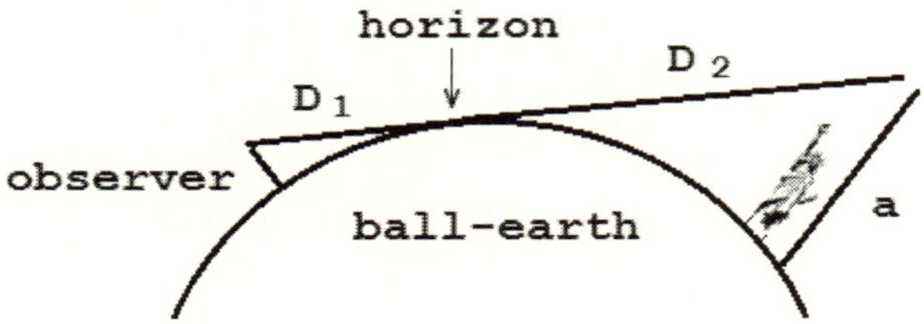

Viewing the statue from out at sea

$observer = 20\, ft$ height of the deck of a ship

$D_1 = \sqrt{\frac{3}{2}(\,20ft\,)} = 5.5\, miles$ from observer to the horizon

$D_2 = \sqrt{\frac{3}{2}(\,326ft\,)} = 22\, miles$ from the horizon to the statue

$D_1 + D_2 = 27.5\, miles$ farthest distance you could view the statue on a ball-earth, on a fictional earth

But on the real earth, on the flat earth
the statue of liberty can be seen from 60 miles away
all the way down to waves crashing on rocks below it

Ball-earth horizon math produces useless numbers
In real life there exists no curve to measure

The camera shows the whole island, top-to-base

Distant buildings are never tilted away behind an edge

When a ship's hull seems to disappear over the horizon
and the mast is still visible
a telescope can bring back into view the entire ship, hull and all

There is no curve whatsoever

It doesn't matter the accuracy of the equation
It's a ball-earth equation
The whole idea of ball-earth is a fantasy
It's not supposed to add up or make sense
The joke's on you

Earth is flat

Midnight Venus

How is it that venus can be viewed at midnight?
Isn't it supposed to be closer to the sun than earth?
And doesn't ball-earth turn away from the sun at night?
Aren't the sightings supposed to occur only at dusk or dawn?

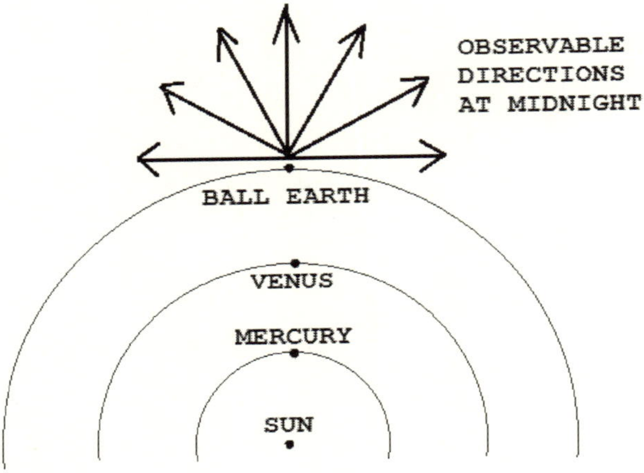

Heliocentric Mercury Venus Earth

If earth is a ball
venus cannot be viewed from earth at midnight

But it does happen
April 2012, Indianapolis, IN (longitude 86°)
Venus sets into the horizon at 12:20 a.m.

(Also from the western flank of any time zone
venus set after midnight)

At what angle would your vision have to dip down
through the ball-earth horizon in order to see venus at 12:20 a.m.?

$12{:}20 \; - \; 1hr \; (daylight \; savings) = 11{:}20 \; p.m.$
$11{:}20 \; p.m. = 80° \; past \; dusk$
$80° - \; 86° \; (Indanapolis) + \; 75° \; (eastern \; standard \; time) = 69°$
69° Global tangent line through Indianapolis as 12:20 a.m.

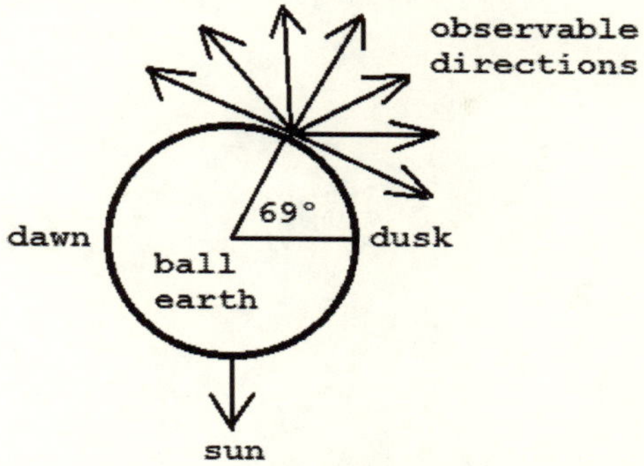

Heliocentric venus orbital radius: 67.24 Million miles away from the sun
Heliocentric earth orbital radius: 92.96 Million miles away from the sun

$\sin(\,46°\,) = \frac{67.24}{92.96}$

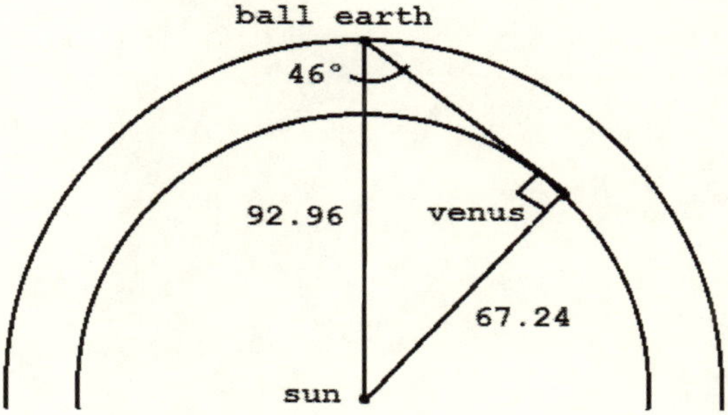

69° – 46° = 23°
Your line of sight has to be angled down 23° below horizontal
through the ball-earth to see venus at 12:20 a.m.

This real-life experience
of watching venus set into the horizon at midnight
is another proof that earth is not a ball

Chapter 4

The Real Flat-earth

The Flat Earth Society is a sabotage organization

They push the misconception of the shape of the flat earth
They plant fallacious imagery into classroom textbooks
Controlled opposition discourages exploration into real flat-earth

You think the flat earth is a disc floating in outer space
You think ships can sail off the edge
You think the sun travels around the underside

On the real flat-earth the sun doesn't go down over the horizon
It goes away into the horizon, farther than its light will reach
Its circular path is the same height above earth all day long

Flat-disk-earth floating in space is a false flag
Those promoting this image are only pretending to be flat-earthers

Floating planetoids of rock and dirt are an illuminati fantasy
Those are just cartoons

Purposefully ridiculous the idea of a disk-earth floating in outer space
Almost as ridiculous as living upside-down on a ball

Real Space

It’s easy relatively for you to imagine the flat earth
when you learn the sun is small and close
when you learn antarctica surrounds the oceans 360°

What seems to be difficult is the truth that outer space is a hoax
There do exist secret space programs
And civilian passengers might someday tour the stars
But it is near space, not outer space
And earth isn’t floating

The stars are close
You can see stars
They are tangible, explorable, achievable

You are trained to imagine flat-earth from a 100,000 mile view
This is an unrealistically distant view
Constant bombardment of ball-earth cartoons
has habitualized you into imagining such great distances

Scientismic priests say the sun’s light takes 8 minutes to reach earth
They fabricate these numbers to make their equations seem plausible
to hide the fact that earth isn’t a ball

They say the light from Betelgeuse takes 430 years to reach earth
But you know you know the stars are alive in real time

Earth is a puddle in the ice

Antarctica surrounds the other continents 360°
Antarctica is the edge of the puddle, an ice wall 150 ft tall

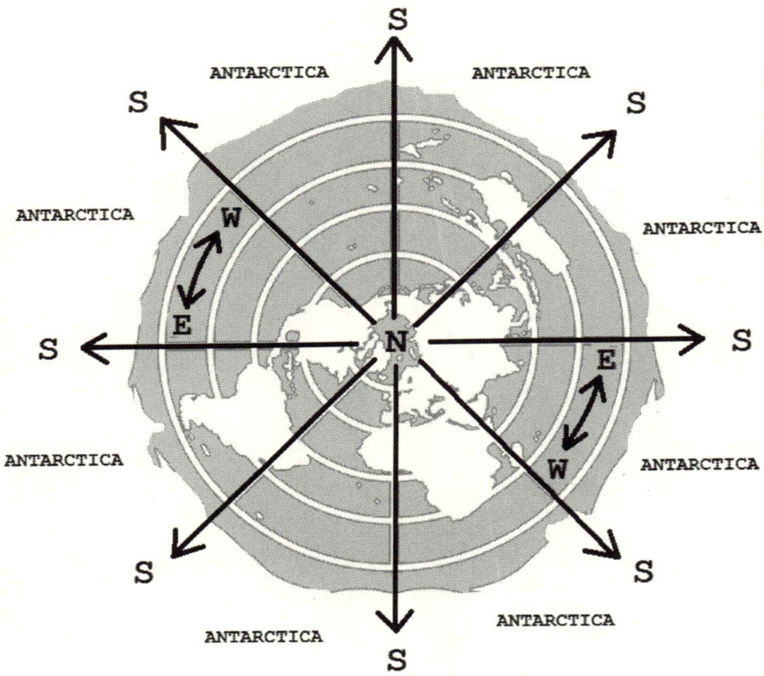

The sun revolves around the center north pole of flat-earth
in horizontal east-to-west circles
Larger circles in the winter, farther south
Smaller circles in the summer, nearer the center north pole

The sun is small and low, near to earth
When the sun is over china it is too far for americans to see

Spherical distance from Cape Town to Dubai: 4,735 miles
Spherical distance from Dubai to Mebourne: 7,231 miles
Spherical distance from Cape Town to Melbourne: 6,416 miles

Oh dear ball-earther, do pray tell
Why would an airline company double the length of the flight path?

If you want to fly from Cape Town to Melbourne
you will layover in the northern "hemisphere"
But why?

Because earth is flat

On a flat-earth map Dubai is right in line
midway between Cape Town and Melbourne

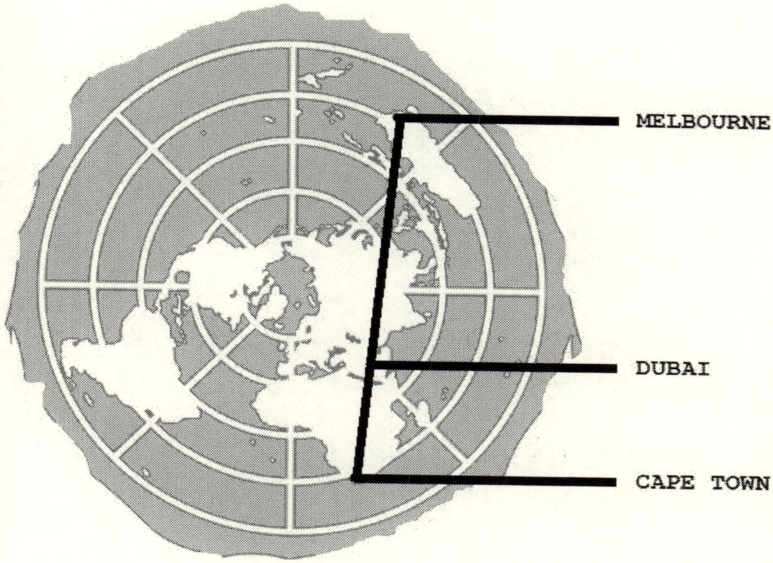

On a non-stop flight from Los Angeles to Malaysia
a woman giving birth caused an emergency landing
IN ALASKA!

Some fake news caster makes some fake excuse
for why the plane was so far off course, spherically speaking

The truth is earth is flat
And the plane was right on course

On a flat-earth map Alaska is right in line
midway between Los Angeles and Malaysia

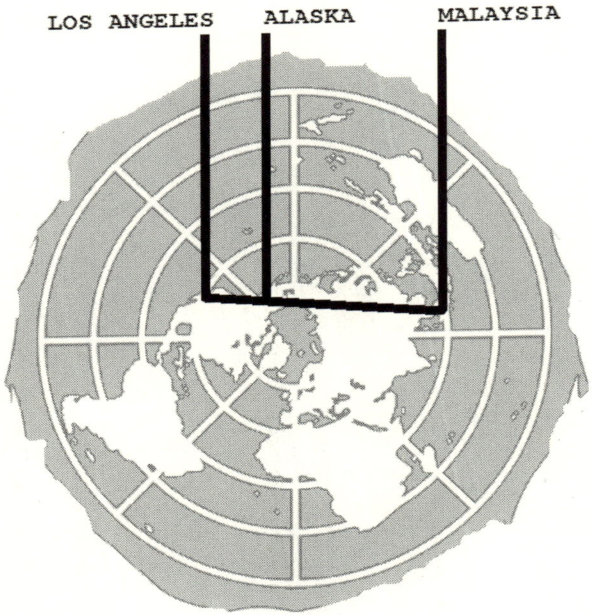

Midnight Sun

In the far north
at the peak of summer
instead of setting, the sun grazes the northern horizon
The sun can be seen all day long, full circle east, south, west, north
And farther north has even better views of the midnight sun

But if earth is a globe, then where's the southern midnight sun?

In the far south in the peak of southern summer there is no midnight sun
They fake it with badly edited videos
along with all the excuses about why they can't show the full 24 hours

In truth the antarctic sun never nears the southern horizon
The antarctic sun does not rise south of due east nor set south of due west
Not even on December 21

Circumnavigating flat-earth is simply an east-west circle
with the compass always pointing to the center magnetic north pole

But curiously, nobody has ever circumnavigated earth north-south

On a flat-earth, north-south circumnavigation is impossible
However on a ball-earth this would be entirely possible
So why hasn't it happened?

Nobody has ever flown straight-over Antarctica
entering in one side and exiting out the other side

No one has ever flown over the south pole
Because there is no south pole
Those are just cartoons

Earth is flat

How high is the sun?

June 21, noon, tropic of cancer, 23.4° latitude
The sun is directly over head, 90° altitude
Viewed from the equator, the sun has 66.6° altitude
Assumption: This altitude angle is well verified

How far is the tropic from the equator?

$$\frac{25{,}000\ miles\ (ball-earth\ circumference)}{360^\circ} = \frac{1{,}632\ miles}{23.4^\circ\ \ (tropical\ latitude)}$$

1,632 *miles* North-south miles from tropic to equator

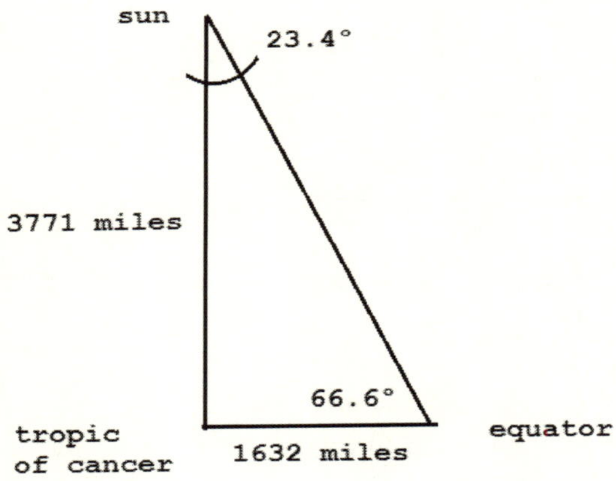

$$1{,}632\ miles * \tan(\ 66.6^\circ\ ) = 3{,}771\ miles$$

3,771 *miles* Vertical height of the sun from flat-earth

So then how big is the sun?

The sun's angular diameter is commonly measured to be 0.533°

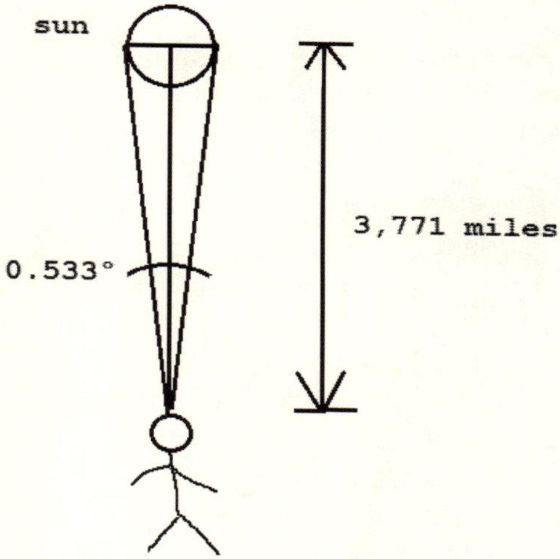

$$sun\ diameter = 2 * 3{,}771 miles * \sin\left(\frac{1}{2} * 0.533° \right)$$

$$sun\ diameter = 35.1\ miles$$

Night time occurs
when the 3771-mile high, 35.1-mile wide sun
travels thousands of miles away to other continents
The sun is too small and low to be seen at such a great distance

How high is the north star?

Polaris is stationary in the sky
night and day and throughout the seasons

Altitude angles derived from spherical math
cannot be trusted to match real astrolabe measurements
especially in the southern regions

Assumption
Real travel-distances between ball-earth latitude lines
are approximately uniform all the way up to the north pole

From my city, latitude 44°, the distance to the north pole is

$$\frac{25{,}000\ miles\ (ball-earth\ circumference)}{360^\circ} = \frac{3{,}194\ miles}{90^\circ - 44^\circ}$$

$3{,}194\ miles$ distance between 44° latitude to the north pole

My astrolabe measurement of the north star is 46°
Slightly off from my latitude

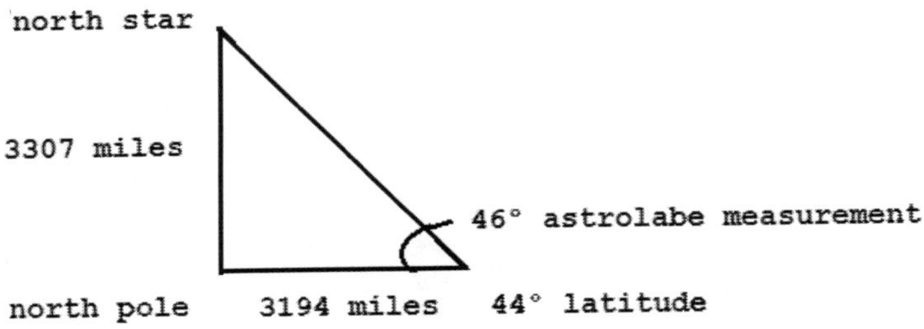

$$3{,}194\ miles * \tan(46^\circ) = 3{,}307\ miles$$

$3{,}307\ miles$ Vertical height of the north star from flat-earth

Stars are Close

From the above calculations
3,307 *miles* Height of the north star from flat-earth
3,771 *miles* Height of the sun from flat-earth

So the north star is lower than the sun?

Maybe the assumption is incorrect that the distance to the north pole can be calculated with the ball-earth latitude lines
(see Appendix C, alternate calculation for north pole distance)

Maybe the assumption is incorrect that the sun's altitude angle viewed from the equator, has been actually measured and verified to be 66.6° north on June 21 and 66.6° south on Dec 21

But these are not wild assumptions
Maybe the sun is higher than all the stars

Mercury and venus supposedly transit the sun

Stars transit the moon as well
Astronomers/photographers have often noted that the moon is transparent since stars can be seen through the moon in the daytime
But that is based on the assumption that stars are farther than the moon

More likely the stars pass in front of the moon

The stars are closer than the moon
So then it makes sense that the stars are closer than the sun

Australia Polaris

Scientismists say it is a myth
that Polaris can be seen from anywhere south of the equator
But their “knowledge” is built on spherical math
which doesn’t apply to the real world

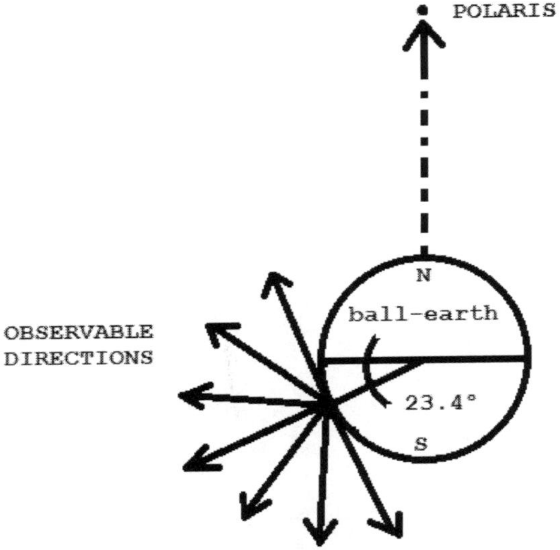

Polaris is viewable from as far south as 23.4° south latitude
according to prominent explorers and military logbooks

From 23.4° south latitude the north star wouldn’t be observable
if earth is a ball

These real-life viewings are proof that earth isn’t a ball

How Big is Antarctica?

The 70° south latitude line
represents a reasonable average antarctic circumference

Judging by a Globe
What is the circumference of a 70° latitude line
on a 25,000 mile circumference ball-earth?

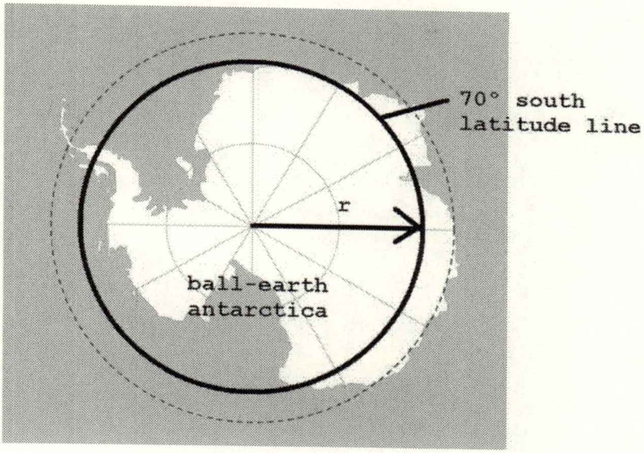

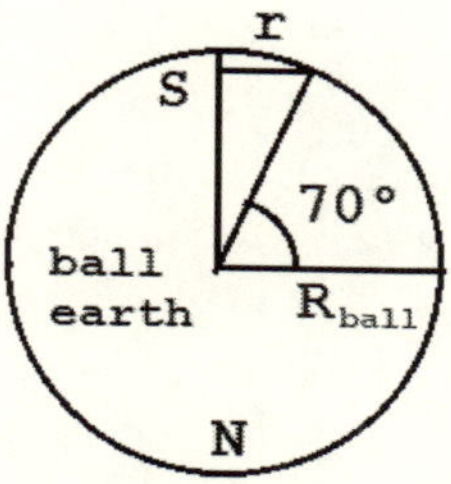

$$r = R_{ball} * \sin(90° - 70°) = 3979\, miles * \sin(20°) = 1361\, miles$$

$$Circumference = 2\pi * 1361\, miles = 8551\, miles$$

Ball-earth circumference of the 70° south latitude line at sea level
A rough estimate of the antarctic coastline length

By comparison what is the circumference of the 70° south latitude line ON FLAT-EARTH?

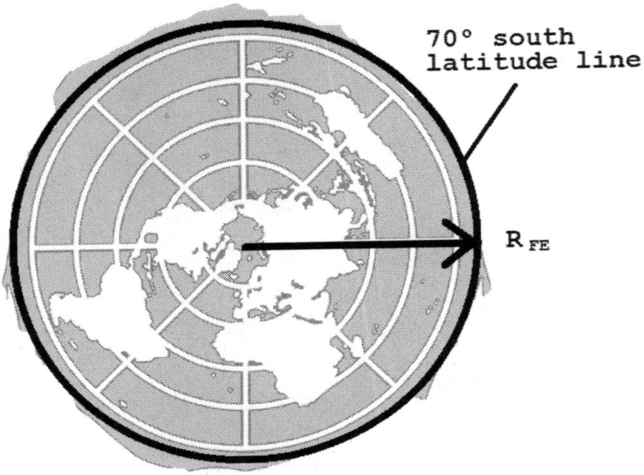

Flat-earth radius of the 70° south latitude line
is a straight line spanning 160° from center north pole to 70° south
$(160° = 90° + 70°)$

$R_{FE} = 160° * \frac{25{,}000\ miles}{360°} = 11{,}111\ miles$
Flat-earth radius at 70° south latitude

$C_{70FE} = 2\pi * 11{,}111\ miles = 69{,}813\ miles$
Flat-earth circumference at 70° south latitude
A rough estimate of the antarctic coastline length

A comparison of antarctic coastlines
69,813 miles (flat-earth) >> 8,551 miles (ball-earth)

There is a frighteningly lot we don't know about antarctica

The Antarctic Treaty

Government permission is difficult to obtain
to allow you to travel any farther south than 60° south latitude

Even if your mission is deemed worthy and you've paid the hefty fees
there are so many environmental restrictions
and all the other national borders to deal with
Access is practically completely prohibited

Sure there are tourist destinations
But no independent exploration

How large the military bases might be
along the tens of thousands of miles
of antarctic coastline never seen by civilians

Why do popes and government officials often visit antarctica?
Who are they meeting with?

The powers that be push the ball-earth idea
to shrink the size of explorable oceans

They restrict public fly-overs of large swaths of sea
to hide their pleasure islands

Chapter 5
Experts

Your priests
Einstein, Newton, Copernicus, Aristotle, nasa et al
They are corrupted to the bones. All of them high-level free masons

All of the astronauts who have flown in space missions are free-masons
Pythagoras was the first free-mason and the first ball-earther

Your priests are generally smart. They ain't wrong
They are lying. They are scamming you

Flapping about strings and relativity and black holes and dark matter
Void of real-world observations
Full of hoaxes and forgeries
They admonish you to believe in the accuracy of cartoons

Defenders of the ball-earth theory love to use terms like
illusion, mirage, unexplained phenomena
to make excuses for why their narrative doesn't jibe with reality

Chicago visible from 60 miles across the water is not a mirage
Lunar eclipses during sun-up hours is not an optical illusion
All these anomalous phenomena prove that earth is not a ball

You are refusing to trust your senses
You put faith in titles and prestige
You have a severe case of expertitis

Obvious

When a fact is obvious scientismists tell you it is false
When a statement is improbable, extraordinary, outrageous
scientismists tell you it is science

Earth is obviously flat
Yet they say it's a ball

Up is obviously up and down is obviously down
Yet they say it depends on your global coordinates

The sun is obviously larger than the stars
Yet they say stars are larger

The sun is obviously the same size as the moon
Yet they say the sun is 400 times larger

Earth is obviously the only plane(t)
Yet they say earth is just one of many floating earthen balls

Earth is obviously stationary
Yet they say it spins and speeds through the solar system

They say earth traverses the milky way 12 Million miles every day
which would obviously cause changes in stellar relative parallax
Yet the stars have had the same patterns for millennia

Moonlight obviously originates from the moon
Yet they say it's reflected sunlight

Spherical surfaces reflect light on only a small area
so the whole moon surface wouldn't be illuminated from reflected light

Reflected sunlight would retain some of the heat properties of sunlight
But focused, concentrated moonlight doesn't heat thermometers at all

Why do you close your mind so tightly?
What makes the ball so worthy of your defense?
Why do you trust the free-masons so lovingly?

Think of how ridiculous the idea: floating earthen spaceballs
You are foolish. You've been fooled
You were but a child. Little children are easy to fool

But you are no longer a child
Now you are ignorant, educationally lobotomized
Unable to judge fact from fiction

You are afraid of being labeled as a thought-criminal
Afraid to admit your trusted authorities force you to believe in untruths
Afraid to realize that the world is actually a dangerous place

You lock onto the question of why
What reason would the authorities have to lie about the world?

But if the motive of a crime is unknown it doesn't invalidate the evidence
The fact is they are lying and you refuse to acknowledge

Truth will free you from the bondage of your experts
The price of freedom is self-responsibility. Too high?

Chapter 6

Science and Religion

A scientist is inquisitive about reality
A scientist is a truth reporter

Science is a torch fending off the encroaching darkness of religion

Science and religion are antithetical
But scientism and religion work hand-in-hand

The vatican infiltrates many fields of science
The vatican owns the world's largest binocular telescope
named Lucifer

Most modern prominent astronomers are catholic priests
The inventor of the big bang theory was a jesuit priest
Copernicus was a catholic priest
Galileo was heavily connected with the church

Theories, fairy tales, dogma, omnipotent priest class
Shame cast upon the non-compliant
Scientism is a cult

Scientism is tired
The studies are too complex/convoluted for a layman
The earth is too large to see the curve
The constellations are too far away to transform
We haven't been back to the moon because been there done that
Evolution is so slow the evidence has disappeared

Don't let them have these excuses
They don't deserve your trust

Evolution is an obvious hoax

Darwinian evolution was initially pushed onto the public scene
by the royal society and free-masonic organizations

The theory makes no logical sense
Close relationships amongst the species
are not evidence that one species is derived from another
and life cannot be spontaneously formed by inanimate materials

The whole idea is ludicrous
And they speak with such arrogant certainty
And they postulate so far back in time
And their fabrications are so twisted
And those with high degrees are generally the most brainwashed

They add the disclaimer that Evolution is just a theory
Yet they treat it as gospel
Professors get fired for questioning the official story
Schoolchildren are force-fed these satanic ideas
Even the vatican promotes evolution

All of Evolution's archeological evidence is hoaxed
The fact that they are hoaxing you
should make you criticize everything they do

Monkey-men are just cartoons

All of the evidence for Evolution has been proven fraudulent

Neanderthal Man
Discovered just in time to coincide with Darwin's book release
has never been shown to be any different than modern humans

Java Man
Concocted from a human leg bone
and a skull cap, jaw fragment, and three teeth of a gibbon

Piltdown Man
Concocted from a human cranial fragment
and a recently deceased orangutan jaw bone

Nebraska Man
Concocted from a single tooth of a pig

Peking Man
Concocted from monkey skull fragments

Orce Man
Concocted from a donkey skull fragment

Turkana Boy
Is a modern human

Lucy
Is a chimpanzee

These are high-profile cases pushed by evolution publications
and proudly displayed in evolution museums
Some are still being promoted

Dinosaurs

Before the 1840's, the concept of dinosaur doesn't exist
The theory was postulated by members of secret societies
before a dinosaur fossil had ever been found

The first dinosaur was the trachodon (toothosaurus)
The whole animal was concocted from just a few teeth

All dinosaur remains have been "discovered" by paleontologists
Not by farmers, construction workers, or people who know the land

Some dinosaur hunters are accredited with the discovery
of hundreds of dinosaur species each

In truth all dinosaur species are fabricated
Dinosaurs never existed
Those are just cartoons

No dinosaur skull has ever been discovered
Their most complete skeleton consists of five bones
The bones could very well be from a giraffe, elephant, whale

The elite paleontologists hold the vault keys
And they don't let real scientists study the bones

All the bones displayed in museums are castings and sculptures
All dinosaur museums and documentaries are funded by free-masons

The dinosaur hoax is a religious combat operation
They are trying to prove evolution and their scientismic religion
They are trying to disprove the bible, the contender

Real Archeology

The bones of giant humanoids have been discovered all over the world
Full skeletons and mummies that are quickly hidden or destroyed
These and other unearthed monsters disprove Darwinian evolution

What about the antediluvian megalithic structures?
Baalbek and the vast tunnel networks of Cusco and Malta
These are constructions that we couldn't replicate today

The walls of sacsayhuaman have been impossibly constructed
The megalithic jigsaw-puzzles are squished together
The results are earthquake-proof

The great pyramids of Giza are not tombs
These gargantuan limestone pyramids have internal granite substructures
and strange passageways and vents and chambers

Did they cut and transport those stones using sonic magic?
Was their technology far beyond our own?

History is clouded
Real discoveries are kept secret

The vatican perches upon many archeological sites
blocking true discovery
They've burned many books and destroyed many cultures

The Smithsonian institute hoards millions of artifacts
deep in their enormous, multi-level warehouses

Those who control the past govern the future

Big Bang

A singular point of nothingness with no space and infinite mass
explodes at an arbitrary time to eventually evolve into everything

Do these "scientists" believe their work relates to reality?
They lecture students about undeniable events
that occurred during the first few trillionths of seconds after the spark

They proudly solve complex math equations
and disregard obvious conflicts with the real world

The Doppler effect as a possible explanation for the Red Shift
is the crux of the whole big bang theory
It has long ago been debunked
The universe isn't expanding
Other real scientists say so too, not just flat-earthers

They say if a star's red shift is so high
then its recession velocity is so fast
If the velocity is so fast
then it must have gotten so far since the big bang
And if it's so far and we can see it with such an optical angle
then it must be so big

This is where they get their numbers, and it's all based on fraud
This is how they tell you that quasars are 13 T-T-TRillion miles away

The Doppler effect connection has been completely shot
The red shift is not a measurement of recession velocity
nor a measurement of distance nor size

When an assumption gets disproven
a real scientist goes back to the drawing-board
Admitting fault is crucial to science

The question of our origin
is only scientific when evidence is involved
Proponents of big bang and evolution are theologists, not scientists

Considering all the hoaxes, you should be seeing red flags
Such hype. Such aversion to examination. Such fraud

There was nothing
Then nothing exploded. This created everything
Condensing nebulous clouds
Earth set revolving the sun
Life randomly evolving from a lifeless molecule
Rocks from "outer space" wiping out the "dinosaurs"

You have to be insane to believe this
Yet you laugh at the truth tellers

Spaceballs and asteroids are cartoons
Dinosaurs and monkey-men are cartoons
Never could the elements align to create life from inanimate material
These are scientismic hoaxes

All this talk of evolution and giants and satanists
causes you to label flat-earthers as religious fanatics
But you've got it backwards

Ball-earthers are the religious ones
Pawns of the most frightening religion

Chapter 7

Aliens

Maybe the existence of aliens can't be completely discounted
There does seem to be some credible stories
of alien abductions and UFO sightings

But there cannot be aliens from stars or planets
Though maybe from beyond the cosmos
or from other dimensions
or from inside earth like the hyperboreans

UFOs are most likely from human dark-government programs
The nazis had anti-gravity flying disc vehicles in the 1940's
This type of technology is channeled from non-human intelligence
For ages oracles have been conveying knowledge from the archons

Alien hype is a deep psychological attack

The groundwork is being laid to stage an alien invasion of earth
The attacking aliens might themselves be man-made
Human-animal hybrid engineering has existed for a long time
DNA can be extracted from the mummies of giants
Who knows what kinds of armies are being grown

They tell you that humans are aliens
They tell you that you are unevolved, incomplete
that superior races will visit earth to teach you a better way

They are preparing you for the second coming of anti-christ/Lucifer

This is actually just their cover story
to unite the nations into an authoritarian one-world government

The globalists will crash the world economies
The anti-christ aliens will offer you cures and food and shelter
But for the price of receiving the mark of the beast

Transhumanism

What is the next step in evolution?
Will humanity join the cult of merging with machines?

Humans have an innate sense of freedom
We do not make very good slaves
And authoritarians hate the road block of civil rights

Perhaps the origin of the visiting superior race will not be outer space
but instead laboratory test tubes
Humanoids spliced from 10 parents
Throw in some genes from a mouse as well
If it's not fully human, then it won't have human rights

However there will be laws against criticizing your new brothers
You will be jailed for hate speech if you say they aren't human

The goal is to replace humans with bio-technologically controlled slaves
to improve the human condition

You are being systematically poisoned
cooked with microwaves
Soon humans will no longer be able to breed outside of laboratories
Those who still can will not be allowed

You are being swept aside
An inferior, out-dated model
Seemingly unable to cope with the rigors of modern living

But alas, you are offered immortality
Upload your mind to the supercomputer
Never mind that your physical body will die in the process
That body was inept, but you can live forever in dataspace

What other than an alien force can hatch such a scheme?

Chapter 8
Spirit

Knowledge about earth's shape is important
It is the world you live upon for God's sake

Stars are more beautiful and intimate
When you realize they are only a few thousand miles away
More attainable

Which stars have been visited by dark government operations?
What is a star's composure?

Where do the tides really come from?
Are there temperate fresh-water oceans around the north pole?
Is this why the arctic ice is fresh-water ice?
Is there a magnetic mountain at the north pole?

To the south, what is beyond the antarctic wall?
Infinite plains of ice? A firmament? An edge? Habitable lands?

What's beyond the firmament?

What is earth? What are you?

Scientismists twist these basic tenants of life
They coerce you into wasting your energies
dreaming their galactic dreams
In reality your only world is your surroundings

Life is all about humans
We are the leading edge in the evolution of consciousness
We are the reason this physical plane exists
We are the creators of experience
We are the riders of time

Masonic overlords scramble time. They mug history
They sever your connection with anything deeper than their lies

Flat-earthers are rapidly growing in number
Thinking-people see through the globalist indoctrination system

The deceivers see the truth coming and they will fight to keep a lid on it
Their societal infiltration is ubiquitous
They position themselves for a new replacement system
a system where truth is outlawed

The whole world is currently engaged in an info war
The globalists/communists are meeting clashes on many fronts
Flat-earth truth is a flanking cavalry attacking the scientismic angle

In the past you've said flat-earth isn't an important issue
But don't you see?
Scientism is intimately involved with every conspiracy

Flat-earth truth is such an obvious weapon against them
It's so easy to prove
yet its system-disruption force is so far-reaching
Truth bulldozes ball-earthers to the side and takes the globalists also

We know they've known about this the whole time
We know they've been suppressing this the whole time
They are all guilty
In one foul swoop, they've all lost their credibility

Instead of fighting the surface battles
flat-earthers scoop the whole war out from underneath and shovel it away

It doesn't have to be a violent overthrow
Anyone who refuses to convert to truth nullifies themselves
inherently, automatically

All we need is enough people on board
Please help spread the message

Appendices

Appendix B

Ball-earth Math: Distance to the Horizon

$a = \frac{2}{3}D^2$

This is not a flat-earth equation, it's a ball-earth equation

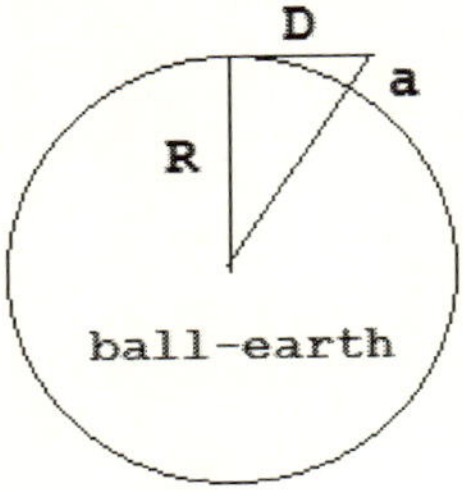

$a = Altitude\ \{\ ft\ \}$
$D = Distance\ \{\ miles\ \}$

$R = R_{ball} = 3979\ miles$

$(R + a)^2 = R^2 + D^2$

$R^2 + 2Ra + a^2 = R^2 + D^2$

$a(2R + a) = D^2$

$a = \frac{1}{2R+a}D^2$

$a \ll R$

$a = \frac{1}{2R}D^2$

$a = \frac{1}{2*(\ 3979\ miles\ )} * D^2 * \frac{5280\ ft}{1\ mile}$

$a = \frac{2}{3}D^2$ The ball-earth horizon is D miles viewed from altitude a feet

Appendix C

If we want to find the distance to the north pole
without relying on arctic ball-earth latitude lines
we can collaborate and triangulate

In textbooks, in planetarium software, in planispheres
altitude angles are calculated from spherical math
Nowhere is there data from actual measurements

Using an astrolabe (a protractor and a string with a weight)
measure the altitude angle of the north star
Aim the straight edge of the protractor and read the string's dangle angle
The horizon is 0° and the zenith is 90°

θ_1 and θ_2 are astrolabe measurements from two distant latitudes
The north-south distance between latitude lines, X_{12}
is more reliable in lands more thoroughly travelled than in the arctic

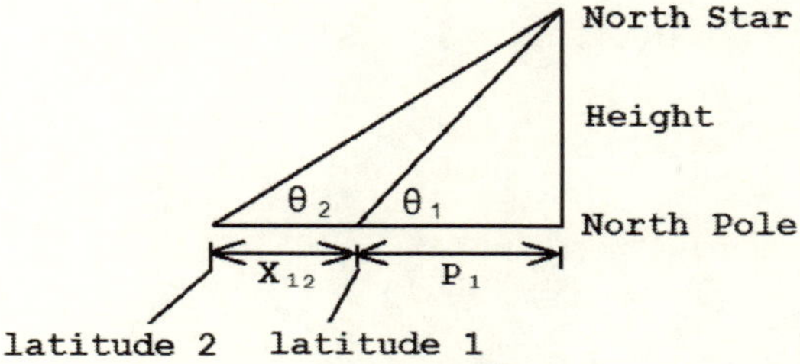

$Height = \frac{X_{12}}{\left(\frac{1}{tan\theta_2} - \frac{1}{tan\theta_1}\right)}$ height of the north star from flat-earth

$P_1 = \frac{X_{12}}{\left(\frac{tan\theta_1}{tan\theta_2} - 1\right)}$ distance to the north pole from latitude 1

$P_2 = \frac{X_{12}}{\left(1 - \frac{tan\theta_2}{tan\theta_1}\right)}$ distance to the north pole from latitude 2

Made in United States
Orlando, FL
04 November 2026

90203414R00037